How Is a Soccer Ball Made?

Angela Royston

Heinemann Library
Chicago, Illinois

Customer Service   888–454–2279

Visit our website at www.heinemannlibrary.com

Photo research by Melissa Allison and
Debra Weatherley
Designed by Jo Hinton-Malivoire and AMR
Printed and bound in China by South China Printing Company

09 08 07 06 05
10 9 8 7 6 5 4 3 2 1

Library of Congress Cataloging-in-Publication Data
Royston, Angela.
  How is a soccer ball made? / Angela Royston.
      p. cm. -- (How are things made?)
Includes bibliographical references and index.
  ISBN 1-4034-6642-4 (library binding - hardcover) -- ISBN 1-4034-6649-1
(pbk.)
1. Footballs--Design and construction--Juvenile literature. I. Title.
II. Series.
GV959.R69 2005
688.7'6352--dc22
                                2004018279

Acknowledgments
The author and publisher are grateful to the following for permission to reproduce copyright material:
Adidas-Salomon AG pp.**5**, **12**, **13**, **14**, **15**, **16**, **17**, **18**, **19**, **20**, **21**, **22**, **23**, **25**; Alamy Royalty-Free p. **28**; AsiaWorks Photography p. **7**; Corbis p. **24** (Walter Hodges); David Wall Photography p.**11**; Getty Images pp. **8/9** (ImageBank), **10** (Stone), **4** (Taxi); Harcourt Education Ltd/Tudor Photography pp. **6**, **26**, **27**, **29**; Science Photo Library p. **28** (Astrid & Hans Frieder Michler).

Cover photograph of soccer balls reproduced with permission of Harcourt Education Ltd/Tudor Photography.

Every effort has been made to contact copyright holders of any material reproduced in this book. Any omissions will be rectified in subsequent printings if notice is given to the publisher.

Some words are shown in bold, **like this.** You can find out what they mean by looking in the glossary.

Contents

What Is in a Soccer Ball?

Many children like to play soccer, especially when they score a goal! Soccer balls are made in a special way so that you can kick them fast and straight.

The cover of the ball has several **layers**.

valve

bladder filled with air

strong **lining**

waterproof casing

A soccer ball is made of **synthetic materials**. These do not wear out even when they are kicked many times. The center of the ball is filled with air.

Who Makes Soccer Balls?

Several **companies** design and sell soccer balls. These companies do not make the balls themselves. They pay factories in other countries to make the balls for them.

This woman is designing new soccer balls.

Factories in China, India, and Pakistan pay their workers much less money than workers in Europe and the United States. This means that the balls cost less to make there.

Many people are needed to make soccer balls.

Where Do Synthetic Materials Come From?

Synthetic materials are made from **oil**. Oil is found deep below the ground. An oil well is drilled into the ground to reach the oil.

Big ships (called oil tankers) or pipes take
the oil to a **refinery**. In the refinery, the oil
is separated into gasoline and other liquids.

Some oil is carried up
from under the seabed.

Making Polyester and PVC

Some oil is made into plastics. There are many different kinds of plastic. **PVC** is a smooth, shiny plastic. PVC is also **waterproof**.

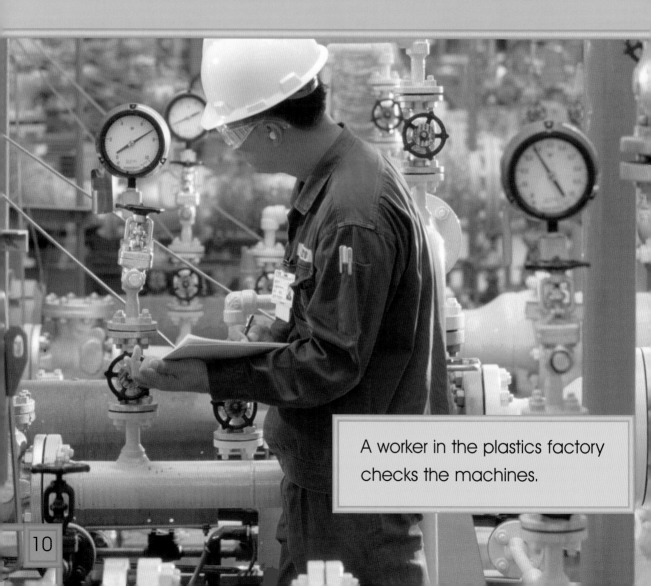

A worker in the plastics factory checks the machines.

Polyester is another kind of plastic. Polyester is made into a strong cloth. Trucks take rolls of PVC and polyester from the plastics factory to the companies that make soccer balls.

PVC

Making the Outer Casing

The outer casing of the soccer ball is made from **PVC**. This helps keep water from getting inside the ball. The **layers** in the **cover** make the ball soft but strong.

Polyester or cotton linings are glued to the PVC.

The outer cover and **linings** are dried until they are firmly stuck together. The **polyester** linings help the ball keep its shape when it is kicked.

Cutting the Panels

A cutting machine cuts the **cover** material into many small panels. Some balls have 26 rectangular panels. Other balls use 24 hexagons and 8 pentagons.

The cutting machine makes little holes around the edge of each panel. The stitches that hold the panels together will go through the holes.

The different panels for each ball are sorted into piles.

Printing the Panels

Each panel is printed. The name of the **company** that designed the soccer ball is printed on some of the panels on each ball.

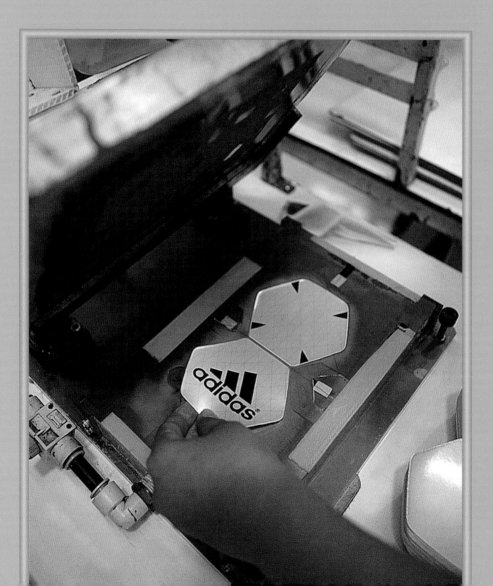

Most companies have a logo. This is a
shape or pattern that the company
uses on the ball along with its name.

Stitching the Panels

Workers stitch the panels together. It takes roughly three hours to stitch the panels to make a soccer ball **cover**.

The workers stitch the panels together with the cover inside out. They leave the last **seam** open. Then they turn the cover with the correct side out and finish the ball from the outside.

Putting the Ball Together

Now the **cover** is ready for the **bladder** to be put inside. The bladder and the cover must be checked carefully.

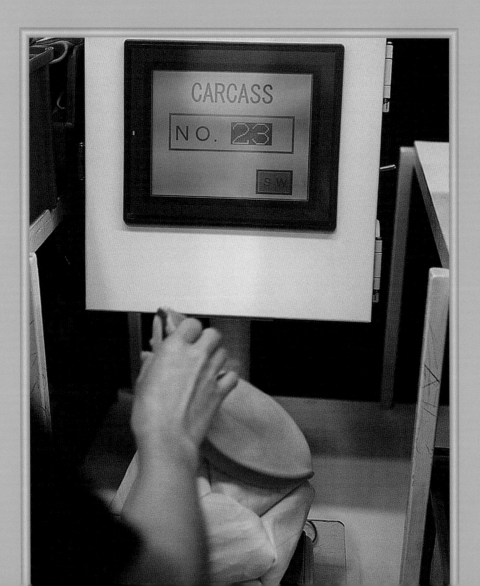

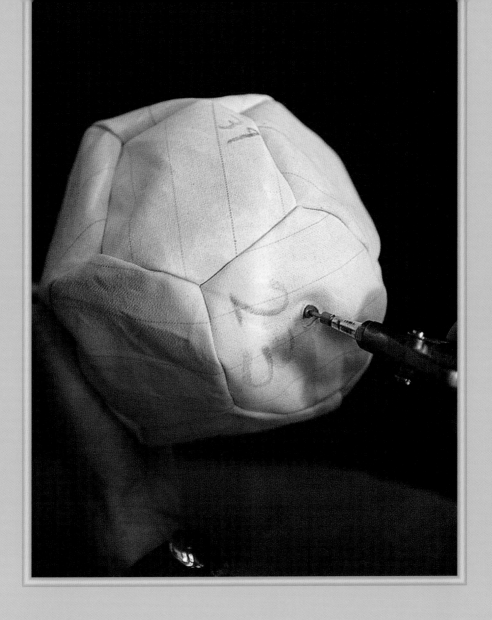

A small hole is cut through the cover and the bladder. The hole is for the **valve**. A special machine puts the valve into the ball.

Finishing the Ball

Air is pumped into the ball to make it round and firm. The **valve** lets air flow into the ball, but stops the air from flowing out again.

If a soccer ball becomes soft, you can pump more air into it.

The air inside the ball makes the ball bounce and fly through the air well. A worker checks the soccer ball to make sure it is the right size.

The **cover** and **bladder** are also weighed before leaving the factory.

Packing and Storing the Balls

Each ball is put inside a clear plastic bag to keep it clean. Trucks and ships take the balls to different countries around the world.

When a store orders soccer balls, the balls are put on a truck and taken to the store.

Selling the Balls

Toy stores and sports stores may sell soccer balls made by several **companies**. You can look at the different balls before you decide which one to buy.

Some of the money you pay for a soccer ball goes to the soccer ball company. The company uses some of this money to make more balls.

From Start to Finish

Soccer balls are made from plastic, which is made from **oil**.

Workers stitch the **cover** together from many small shapes.

The **bladder** is put inside the casing.

Air is pumped into the bladder to make the ball firm and round.

A Closer Look

Soccer ball covers are made with a range of shapes. The shapes can be rectangles, pentagons, or hexagons.

hexagon

pentagon

rectangle

Glossary

bladder stretchy bag that holds air or liquid

company group of people who work together

cover outer layer

layer single thickness

lining material that covers the inside of another material

oil liquid that forms under the ground

polyester type of fabric made from plastic

PVC tough, bendy type of plastic

refinery place where oil is separated into gasoline and other liquids

seam row of stitches

synthetic material material made from plastics or coal

valve gadget that allows something such as air to flow in one direction only

warehouse building where things are stored

waterproof not letting water soak through

More Books to Read

Clark, Brooks. *Kids' Book of Soccer: Skills, Strategies, and the Rules of the Game*. New York: Citadel, 1997.

Crisfield, Deborah. *The Everything Kids' Soccer Book*. Avon, Mass.: Adams Media, 2002.

Gibbons, Gail. *My Soccer Book*. New York: Harper Collins, 2000.

Index